"Shhhhh!" whispered Charlie. He and Brian were creeping through the jungle.

Then Charlie said softly, "I see a cat up ahead. Maybe it could be a lion or a tiger."

1

Charlie and Brian peered through the long grass. They got closer to the cat.

"Do you notice anything?" asked Charlie.

"This baby cat has no stripes," said Brian.

"It must be a lion cub," said Charlie. "We cannot take a lion cub from its mother. That would be cruel!"

Then the boys heard some strange sounds.

"What is that?" asked Brian.

"Maybe it is a fox," said Charlie.

"Do you think it could be a wolverine?" asked Brian. "I saw a picture of a wolverine once!"

Charlie said, "I'll look. You stay here."
Charlie snuck through the thick
green undergrowth. "It is not what you
imagine," Charlie called over to Brian.
"This animal won't bother us."

The two explorers went around the back of a shed.

"We need to be more mobile," said Brian. "Can this machine help us move faster?"

Charlie and Brian went to examine the machine. Charlie said, "This machine is too old and fragile. And it would require gasoline."

"That is too bad," said Brian. "We must find a safe place to rest."

"Come with me," said Charlie.

"Where are we going?" asked Brian. "I am too hot!"

"We'll be there soon," said Charlie. "I know a medicine woman. I am sure she'll have something for us."

Soon Charlie and Brian came to the medicine woman's office.

"Medicine Woman, can you help us? We are hot and thirsty," said Charlie.

Medicine Woman said, "I think you need fluids. Drink this."

Charlie and Brian drank the fluids she gave them. The snacks were also good.